U.S. SPECIAL OPS FORCES

INSIDE THE GREEN BERETS

HOWARD PHILLIPS

PowerKiDS press

NEW YORK

Published in 2022 by The Rosen Publishing Group, Inc.
29 East 21st Street, New York, NY 10010

First Edition

Editor: Greg Roza
Designer: Rachel Rising

Portions of this work were originally authored by Drew Nelson and published as *Green Berets*. All new material in this edition was authored by Howard Phillips.

Photo Credits: Cover South_agency/E+/Getty Images; cover, pp. 1–32 CRVL/Shutterstock.com; cover, p. 1 Zsschreiner/Shutterstock.com; cover, pp. 1, 3, 30, 31, 32 Massimo Saivezzo/Shutterstock.com; pp. 5, 7, 9, 11, 13, 19, 21, 23, 29 Khvost/Shutterstock.com; p. 5 https://commons.wikimedia.org/wiki/File:Seven_Green_berets.jpg; p. 6 https://commons.wikimedia.org/wiki/File:Merrill%27s_Marauders1.jpg; p. 7 https://commons.wikimedia.org/wiki/File:1st_Special_Service_Force_members_being_briefed_at_Anzio_3396066.jpg; p. 9 Ralph Crane/Contributor/The LIFE Picture Collection/Getty Images; p. 11 https://commons.wikimedia.org/wiki/File:JFKRiceUniversity.jpg; p. 13 Larry Burrows/Contributor/The LIFE Picture Collection/Getty Images; p. 15 Sunset Boulevard/Contributor/Corbis Historical/Getty Images; p. 17 alessandro guerriero/Shutterstock.com; p. 19 https://commons.wikimedia.org/wiki/File:Fort_Bragg_1st_Brigade_barracks.jpg; p. 21 USAF/Handout/Getty Images News/Getty Images; p. 23 Remi Benali/Contributor/Hulton Archive/Getty Images; pp. 25, 29 Joe Raedle/Staff/Getty Images News/Getty Images; p. 27 ROBERT ATANASOVSKI/Contributor/AFP/Getty Images.

Library of Congress Cataloging-in-Publication Data

Names: Phillips, Howard, 1971- author
Title: Inside the Green Berets / Howard Phillips.
Description: New York : PowerKids Press, [2022] | Series: U.S. Special Ops forces | Includes index.
Identifiers: LCCN 2020045460 | ISBN 9781725328952 (library binding) | ISBN 9781725328938 (paperback) | ISBN 9781725328945 (6 pack)
Subjects: LCSH: United States. Army. Special Forces–Juvenile literature. | Special forces (Military science)–United States–Juvenile literature.
Classification: LCC UA34.S64 P544 2022 | DDC 356/.1670973–dc23
LC record available at https://lccn.loc.gov/2020045460

Manufactured in the United States of America

CPSIA Compliance Information: Batch #BSPK22. For further information contact Rosen Publishing, New York, New York at 1-800-237-9932.

CONTENTS

U.S. ARMY SPECIAL FORCES

The job of the U.S. Army Special Forces is to keep Americans safe. They carry out many different kinds of missions to make sure that happens. They fight enemies, but they also do other secret activities and spread goodwill to other nations.

The U.S. Army Special Forces have a nickname: The Green Berets. The name comes from the green hats that are an official part of their uniforms. They fight and work all over the world, during both times of war and times of peace. Green Berets are specially trained soldiers who work in small groups to carry out **covert** missions too difficult or dangerous for regular military units.

SPECIAL FORCES IN ACTION

THE GREEN BERETS' TRAINING IS VERY PHYSICAL AND INCLUDES COMBAT, **RECONNAISSANCE**, AND OTHER TYPICAL MILITARY ACTIVITIES. HOWEVER, THEY'RE ALSO TRAINED IN WORLD POLITICS, **DIPLOMACY**, LANGUAGES, AND CULTURES. THIS ALLOWS THEM TO WORK WITH PEOPLE IN FOREIGN COUNTRIES TO HELP THEM. GREEN BERETS ARE ALSO TRAINED IN LESS COMMON MILITARY ACTIVITIES, SUCH AS SPREADING FALSE INFORMATION, TO MAKE IT HARDER FOR ENEMY FORCES TO OPERATE.

AS EARLY AS 1953, GREEN BERETS WERE USED AS UNOFFICIAL HEADGEAR FOR SPECIAL FORCES SOLDIERS WHEN THEY WERE IN THE FIELD. DURING THE EARLY 1960S, THE DEPARTMENT OF THE ARMY MADE IT THE OFFICIAL HEADGEAR OF THE U.S. ARMY SPECIAL FORCES.

BEFORE THE GREEN BERETS

The Green Berets weren't the first group of Special Forces in the U.S. Army. During World War II, the army created the Office of Strategic Services (OSS) to collect secret information about the enemy. They traveled into enemy territory to help local resistance groups in their secret struggle against their enemies.

In 1942, a group of Canadian and American soldiers in Montana formed the 1st Special Service Force. They were trained in mountain climbing and in air and water operations. They fought in the mountains of Italy and in the waters around southern France. They were nicknamed "The Devil's Brigade."

THIS PHOTOGRAPH FROM APRIL 1944 SHOWS MEMBERS OF THE 1ST SPECIAL SERVICE FORCE GATHERING FOR ORDERS DURING THE BATTLE OF ANZIO IN ITALY.

THE FIRST GREEN BERETS

The Army Rangers were a Special Forces group that fought during the Korean War. After the war, the Rangers were shut down, opening the way for a new Army Special Forces unit. In 1952, the army gave the new 10th Special Forces Group 2,300 spots to fill with specially trained soldiers—the first Green Berets.

Colonel Aaron Bank was the group's first commander. He believed the United States needed a group ready for "unconventional warfare." This term may mean different things to different people, but it often involves actions different from typical combat and commonly includes covert missions in enemy territory.

THE 10TH SPECIAL FORCES GROUP

WHEN THE 10TH SPECIAL FORCES GROUP WAS FIRST FORMED, IT INCLUDED COLONEL AARON BANK, ONE OTHER OFFICER, AND EIGHT ENLISTED SOLDIERS. WITHIN A FEW MONTHS, HOWEVER, SOLDIERS STARTED TRAINING FOR THE GROUP BY THE HUNDREDS. THE UNIT QUICKLY GREW INTO A STRONG, UNIFIED TEAM.

THE FIRST SPECIAL FORCES GROUP WAS ESTABLISHED IN MAY 1952 AT FORT BRAGG, NORTH CAROLINA, BY THE SPECIAL OPERATIONS DIVISION OF THE **PSYCHOLOGICAL WARFARE** CENTER. TODAY, THIS TRAINING SCHOOL IS CALLED U.S. ARMY JOHN F. KENNEDY SPECIAL WARFARE CENTER AND SCHOOL (SWCS).

THE KOREAN WAR

The Green Berets had grown into an important part of the U.S. Army by the end of 1952. A few members of the 10th Special Forces Group became some of the first Army Special Forces troops to be deployed behind enemy lines. They were placed on islands along the North Korean coast and trained local fighters, called "Donkeys" or "Wolfpacks," to raid enemy camps and rescue downed airmen.

Guided by these Special Forces, the "Wolfpack" fighters eventually grew to more than 22,000 men. They were very influential in bringing an end to the Korean War.

IN APRIL 1962, PRESIDENT JOHN F. KENNEDY SAID THE GREEN BERETS WERE "A SYMBOL OF EXCELLENCE, A BADGE OF COURAGE, A MARK OF **DISTINCTION** IN THE FIGHT FOR FREEDOM."

DONKEYS AND WOLFPACKS

THE FIRST ACTION OF THE GREEN BERETS DURING THE KOREAN WAR WAS TO TRAIN LOCAL SOUTH KOREAN FORCES IN **GUERILLA WARFARE**. IN 1952, THEY TRAINED APPROXIMATELY 12,000 "DONKEY" SOLDIERS, AND APPROXIMATELY 3,800 "WOLFPACK" FORCES. MAJOR RICHARD M. RIPLEY, A GREEN BERET COMMANDER, SAID, "OUR MISSION WAS TO HARASS AND **INTERDICT** THE REAR AREAS. WE CONDUCTED RAIDS AND **AMBUSHES** AND LAID MINES ALONG THE MAIN SUPPORT ROUTES."

THE WAR IN VIETNAM

The Green Berets played an important role throughout the Vietnam War. In fact, the first and last soldiers killed during the war were both Green Berets.

The Green Berets were first deployed to South Vietnam in June 1957 to train local fighters in **counterinsurgency**. More Green Berets arrived during the late 1950s and early 1960s. They trained the people of Vietnam to become fighting forces. They also built schools and hospitals, and dug canals to help the people of South Vietnam. The Green Berets left Vietnam in 1971. A total of 17 Medals of Honor were awarded to Green Berets who had fought in Vietnam.

DE OPPRESSO LIBER

THE MOTTO OF THE GREEN BERETS IS *DE OPPRESSO LIBER*. THIS IS LATIN FOR "TO FREE THE OPPRESSED." GREEN BERETS ARE TRAINED TO HELP PEOPLE IN NEED, WHETHER THEY'RE U.S. CITIZENS OR NOT.

IN THIS PHOTOGRAPH FROM 1964, A U.S. SPECIAL FORCES CAPTAIN SPEAKS WITH HEADQUARTERS DURING A RAID ON AN ENEMY LOCATION IN VIETNAM.

POPULAR CULTURE

Green Beret mission are top secret. Stories about the military unit have long interested the American people. The Green Berets have made many appearances in music and movies and even as toys.

In 1966, former Special Forces Staff Sergeant Barry Sadler had a hit in the United States with the song "Ballad of the Green Berets." It was released on an album of songs about being a soldier during the Vietnam War. Several popular film characters, including John Rambo and Jason Bourne, were supposed to be former U.S. Army Special Forces soldiers. Many G.I. Joe action figures were made in the late 1960s to look like Green Berets.

BETWEEN 1982 AND 2019, ACTOR SYLVESTER STALLONE PLAYED FORMER GREEN BERET JOHN RAMBO IN FIVE ACTION MOVIES.

SPECIAL MISSION

The Green Berets are trained to carry out several specific kinds of missions. In direct action missions, Green Berets quickly attack enemy targets to take or destroy enemy supplies, or recover fellow soldiers. Special reconnaissance is used to sneak into enemy locations to gather intelligence. In counterterrorism missions, Green Berets follow the actions of terrorist groups and act directly against them to prevent terrorist attacks. Green Berets use guerrilla warfare to fight with forces in enemy-held territory.

Training others is an important part of being a Green Beret. During foreign internal defense missions, they train and help soldiers in other countries respond to threats from common enemies.

THE GREEN BERETS HAVE A WIDE VARIETY OF EQUIPMENT. THIS INCLUDES PARACHUTES FOR JUMPING OUT OF PLANES, NIGHT-VISION GOGGLES FOR NIGHTTIME MISSIONS, AND TOOLS FOR MOUNTAIN CLIMBING.

GREEN BERET DUTY STATIONS

Army Special Forces Teams have duty stations in different locations around the country. The 1st Special Forces Group (SFG) is at Joint Base Lewis-McChord in Washington State. The 3rd SFG is located at Fort Bragg in North Carolina. The 5th SFG is located at Fort Campbell in Kentucky. The 7th SFG is located at Elgin Air Force Base in Florida, and also has a unit in Puerto Rico. The 10th SFG is at Fort Carson in Colorado.

Green Berets are also deployed to foreign locations so they can reach destinations faster if an emergency occurs in a different part of the world. These include Okinawa, Japan; and Stuttgart, Germany.

BRONZE BRUCE

THE SPECIAL WARFARE MEMORIAL STATUE, POPULARLY KNOWN AS "BRONZE BRUCE," HAS STOOD AT FORT BRAGG SINCE NOVEMBER 1969. IT WAS THE FIRST VIETNAM WAR MEMORIAL IN THE UNITED STATES. THE STATUE IS 22 FEET (6.7 M) TALL. THE BASE IS MADE OF GREEN GRANITE. INSIDE THE BASE IS A GREEN BERET AND A SPECIAL FORCES UNIFORM.

THE U.S. ARMY SPECIAL FORCES TRAINING SCHOOL AT FORT BRAGG, IN FAYETTEVILLE, NORTH CAROLINA, WAS ESTABLISHED IN 1918. AFTER THE PSYCHOLOGICAL WARFARE CENTER WAS OPENED THERE IN 1952, IT BECAME THE CENTER FOR THE GREEN BERETS.

BECOMING A GREEN BERET

If you want to be a Green Beret, you'll need to undergo one of the most difficult training programs in the U.S. military. After basic training, soldiers must go to a 6-week Special Forces Preparations Course (SFPC) at Fort Bragg to prepare for the physical tests in the rest of their training. But even after this course, a soldier still may not qualify to become a Green Beret.

After the preparation course, soldiers take part in the Special Forces Assessment and Selection (SFAS) at Camp Mackall, North Carolina. This 24-day program tests soldiers' mental and physical endurance. The training includes marches, runs, and survival exercises.

SFAS CAMP

THE SPECIAL FORCES ASSESSMENT AND SELECTION CAMP IS HELD FOUR TIMES A YEAR, WITH AROUND 300 SOLDIERS ATTEMPTING THE PROGRAM EACH TIME. DURING THE COURSE, THE CANDIDATES ARE KEPT AWAKE FOR A LONG TIME AND PUT UNDER MENTAL AND PHYSICAL PRESSURE. BY THE SECOND WEEK, MORE THAN HALF OF THE CANDIDATES QUIT THE PROGRAM OR ARE REMOVED BY INSTRUCTORS.

SOLDIERS TRAINING TO BECOME GREEN BERETS MUST WORK TOGETHER TO COMPLETE TEAM ACTIVITIES.

Upon passing the SFAS, candidates then move on to the Special Forces Qualification Course (SFQC). This six-step program prepares soldiers to become Green Berets. Soldiers work on individual skills, such as land **navigation** and small-unit tactics. Next, they complete Military Occupation Specialty training, in which each soldier learns different skills based on his background and interests.

The next step is collective training, where soldiers learn to work in smaller groups. After this, all candidates take part in language training, where they learn at least one other language. The last step is to take the SERE course. SERE stands for Survival, Evasion, Resistance, and Escape.

SURVIVAL SKILLS ARE AN ESSENTIAL COMPONENT TO GREEN BERET TRAINING. HERE, ONE CANDIDATE IS EATING A GRASSHOPPER, WHICH IS A GREAT SOURCE OF NUTRITION WHEN OTHER FOODS AREN'T AVAILABLE.

DO YOU HAVE WHAT IT TAKES?

EACH CANDIDATE FOR THE GREEN BERETS MUST MEET CERTAIN QUALIFICATIONS. THEY NEED TO BE A U.S. CITIZEN BETWEEN THE AGES OF 20 AND 32. THEY ALSO NEED TO BE ACTIVE SOLDIERS WITH AIRBORNE TRAINING. CANDIDATES NEED TO BE ABLE TO SWIM IN FULL GEAR, DO AT LEAST 49 PUSHUPS, 59 SIT-UPS, AND COMPLETE A TWO-MILE RUN IN 15 MINUTES AND 12 SECONDS.

FOREIGN INTERNAL DEFENSE

The Green Berets are often called upon during peacetime to aid in foreign internal defense. They help people all over the world. The Green Berets work with many nations' militaries and police forces on various skills, human rights issues, and **humanitarian** projects.

In order to do this, most Green Berets go through Live Environment Training. Here, they receive training that puts them completely into a new society. In addition to learning to speak the country's language, Green Berets learn the customs and traditions of that country so they can live like a citizen. This training allows them to better help people in foreign countries.

AFTER A MASSIVE EARTHQUAKE HIT THE ISLAND COUNTRY OF HAITI IN 2010, U.S. ARMY SPECIAL FORCES RUSHED THERE TO OFFER SUPPORT TO THE HAITIAN PEOPLE.

THE WAR ON TERROR

The U.S. Army Special Forces have played a large role in U.S. military actions since the terrorist attacks on the United States in 2001. The U.S. military started Operation Enduring Freedom in Afghanistan to fight the terrorists responsible for the attacks. The 5th Special Forces Group joined a group known as Task Force Dagger and helped control operations there.

Starting in October 2001, the 3rd and 7th Special Forces Groups helped train and fight with the Afghan army and police forces. The Green Berets also worked with local tribes to fight terrorists and find those who had gone into hiding.

U.S. ARMY SPECIAL OPERATIONS SOLDIERS TAKE PART IN AN INTERNATIONAL TRAINING EXERCISE IN KRIVOLAK, MACEDONIA, IN 2019.

INTO THE FUTURE

The United States started Operation Iraqi Freedom in March, 2003, to overthrow the government of Iraq. The 5th Green Beret group was there soon after to prevent the Iraqis from firing long-range bombs. The 10th group operated Task Force Viking to train, arm, and fight with local military and citizens to secure the country's borders.

The Green Berets will continue to carry out special missions, both fighting as soldiers and helping to support other friendly nations. Above all else, their job is and always will be to operate all over the world and free the oppressed wherever they go.

U.S. SPECIAL FORCES HAVE FORMED MILITARY UNITS WITH SOLDIERS IN OTHER COUNTRIES, SUCH AS IRAQ (SHOWN HERE), IN A JOINT EFFORT TO BATTLE TERRORISTS AROUND THE WORLD.

THE FIRST FEMALE GREEN BERET

UP UNTIL 2020, ONLY MEN HAD BECOME GREEN BERETS. HOWEVER, THAT YEAR A FEMALE NATIONAL GUARD SOLDIER GRADUATED FROM ARMY SPECIAL FORCES TRAINING. SHE BECAME THE FIRST WOMAN TO EARN THE TITLE OF GREEN BERET! AS OF THE WRITING OF THIS BOOK, THE FEMALE GREEN BERET'S NAME HAS NOT BEEN ANNOUNCED TO THE PUBLIC FOR SECURITY REASONS.

GLOSSARY

ambush: To attack an enemy by surprise from a hiding spot.

counterinsurgency: Organized military action against a group trying to overthrow a government.

covert: Made or done secretly.

diplomacy: The work of keeping good relations between the governments and peoples of different countries.

distinction: Something that makes a person or thing special or different.

guerrilla warfare: Having to do with methods of war that include covert operations, such as small teams, reconnaissance, and secret attacks against an enemy.

humanitarian: A person who works to make other people's lives better.

interdict: To stop or cut off an enemy group by force.

navigation: The science of figuring out the position and course of a ship or aircraft.

psychological warfare: Having to do with methods of war that are done to make an enemy feel less confident, hopeless, or afraid.

reconnaissance: Military activity in which soldiers are sent to find out information about an enemy.

FOR MORE INFORMATION

BOOKS

Rose, Simon. *Green Berets*. New York, NY: AV2 by Weigl, 2014.

Terrell, Brandon. *Missions of the U.S. Green Berets*. Troy, Michigan: Momentum Books, 2016.

WEBSITES

How the Green Berets Work

science.howstuffworks.com/green-beret.htm

Read more about how the Green Berets work at this informative website.

U.S. Army Special Forces

www.goarmy.com/special-forces.html

Visit the official website for U.S. Army Special Forces to learn more about the Green Berets.

Publisher's note to educators and parents: Our editors have carefully reviewed these websites to ensure that they are suitable for students. Many websites change frequently, however, and we cannot guarantee that a site's future contents will continue to meet our high standards of quality and educational value. Be advised that students should be closely supervised whenever they access the internet.

INDEX